HARCOURT

Math
GEORGIA EDITION

INTERVENTION
Problem Solving

Grade 1

Harcourt

Visit The Learning Site!
www.harcourtschool.com

Copyright © by Harcourt, Inc.

All rights reserved. No part of this publication may be reproduced or transmitted in any form or by any means, electronic or mechanical, including photocopy, recording, or any information storage and retrieval system, without permission in writing from the publisher.

Permission is hereby granted to individuals using the corresponding student's textbook or kit as the major vehicle for regular classroom instruction to photocopy Copying Masters from this publication in classroom quantities for instructional use and not for resale. Requests for information on other matters regarding duplication of this work should be addressed to School Permissions and Copyrights, Harcourt, Inc., 6277 Sea Harbor Drive, Orlando, Florida 32887-6777. Fax: 407-345-2418.

HARCOURT and the Harcourt Logo are trademarks of Harcourt, Inc., registered in the United States of America and/or other jurisdictions.

Printed in the United States of America

ISBN 13: 978-0-15-349617-2
ISBN 10: 0-15-349617-7

If you have received these materials as examination copies free of charge, Harcourt School Publishers retains title to the materials and they may not be resold. Resale of examination copies is strictly prohibited and is illegal.

Possession of this publication in print format does not entitle users to convert this publication, or any portion of it, into electronic format.

4 5 6 7 8 9 10 11 0956 15 14 13 12 11 10 09

CONTENTS

- **Using Math Maps** .... **IPS**v
- **Math Map–Join** .... **IPS**vi
- **Math Map–Separate** .... **IPS**vii
- **Math Map–Compare** .... **IPS**viii
- **Math Map–Part-Part-Whole** .... **IPS**ix
- **Math Map–Equal Groups** .... **IPS**x
- **Problem Solving Think Along (written)** .... **IPS**xi
- **Problem Solving Think Along (oral)** .... **IPS**xii
- **Strategies/Skills**
  - 1 Make a Picture **IPS**1
  - 2 Make a Picture **IPS**3
  - 3 Make a Picture **IPS**5
  - 4 Use Objects **IPS**7
  - 5 Make a Picture **IPS**9
  - 6 Choose the Operation **IPS**11
  - 7 Make a Graph **IPS**13
  - 8 Use Objects **IPS**15
  - 9 Use a Model **IPS**17
  - 10 Use Objects **IPS**19
  - 11 Choose the Operation **IPS**21
  - 12 Choose the Operation **IPS**23
  - 13 Use Objects **IPS**25
  - 14 Use Data from a Table **IPS**27
  - 15 Choose a Method **IPS**29
  - 16 Use Objects **IPS**31

Intervention • Problem Solving

17	Act It Out	**IPS33**
18	Act It Out	**IPS35**
19	Act It Out	**IPS37**
20	Use Data from a Graph	**IPS39**
21	Make Reasonable Estimates	**IPS41**
22	Use Objects	**IPS43**
23	Choose the Measuring Tool	**IPS45**
24	Choose a Method	**IPS47**

▶ **Problem Bank** **IPS49**

Using Math Maps

Use an Intervention • Problem Solving Math Map to

- see how the parts of a word problem fit together.

- show what you know and what you are to find.

- write a number sentence or equation to represent a problem.

- solve a problem.

Marcus has 13 stickers, and Juan has 6 stickers. How many more stickers does Marcus have than Juan?

Larger Amount
13 stickers

Smaller Amount
6 stickers

Difference
?

Join

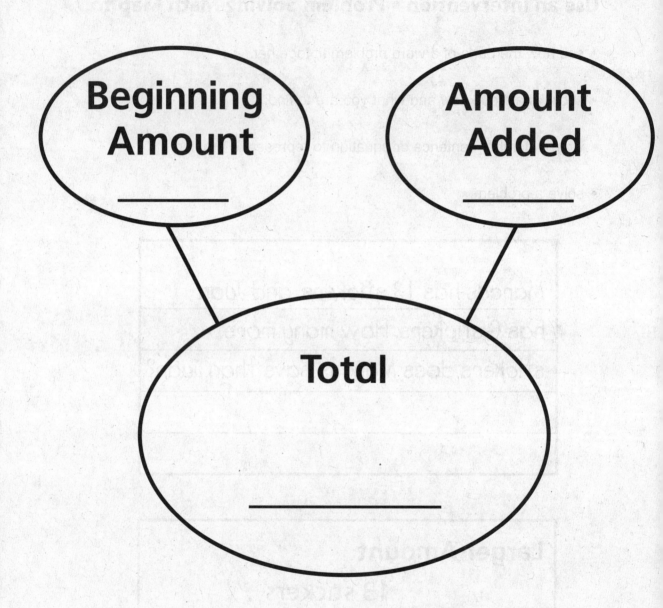

Add or Subtract

Separate

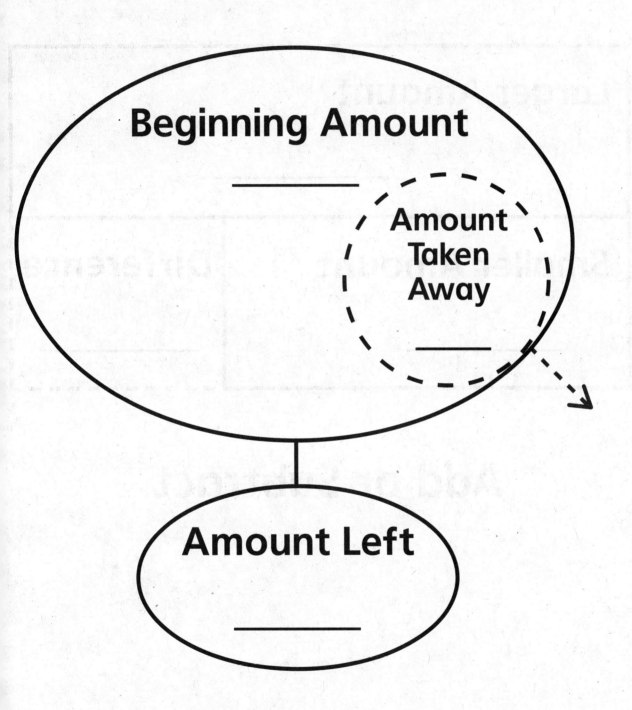

Add or Subtract

Compare

| Larger Amount _____ | |
| Smaller Amount _____ | Difference _____ |

Add or Subtract

Part-Part-Whole

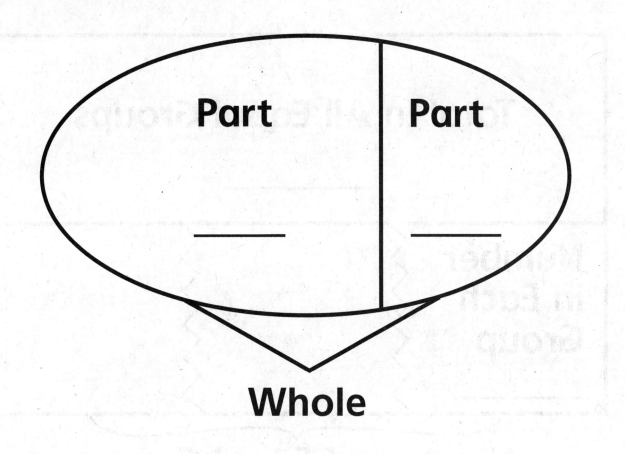

Add or Subtract

Intervention • Problem Solving IPSix

Equal Groups

Total In All Equal Groups
_____

Number in Each Group
_____

Number of Equal Groups

_____

Multiply or Divide

IPSx Intervention • Problem Solving

Name _____

PROBLEM SOLVING THINK ALONG

Problem Solving

Understand

1. Tell the problem in your own words.
   _____
   _____
   _____

2. What do you want to find out?
   _____
   _____
   _____

Plan

3. How will you solve the problem?
   _____
   _____
   _____

Solve

4. Show how you solved the problem.
   _____
   _____
   _____

Check

5. How can you check your answer?
   _____
   _____
   _____

Problem Solving Think Along (written)   Intervention • Problem Solving **IPSxi**

Name _____

PROBLEM SOLVING THINK ALONG

Problem Solving

Understand

1. What is the problem about?

2. What is the question?

3. What information is given in the problem?

Plan

4. What problem solving strategies might I try to help me solve the problem?

5. About what do I think my answer will be?

Solve

6. How can I solve the problem?

7. How can I state my answer in a complete sentence?

Check

8. How do I know whether my answer is reasonable?

9. Did I answer the question?

10. How else might I have solved the problem?

Name _____   Use with Lesson 1.10.

Problem Solving Strategy 1: Make a Picture

How much do you spend for both?

UNDERSTAND

Tell what you know.   A jet costs _____ ¢.
                     A rocket costs _____ ¢.

PLAN

Make a picture to help you solve the problem.

First draw 4 . Next draw 2 .

SOLVE

How many there are in all?

So, you spend __6__ ¢ for both.

CHECK

• Explain why you think your answer is right.

▶ **Try These** ....

Use to show each price.

Draw the (1¢). Write how many there are in all.

1. How much do you pay for both?

_____ ¢

Intervention • Problem Solving **IPS1**

Name _____

PRACTICE ON YOUR OWN

Use to show each price.

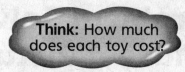
Think: How much does each toy cost?

Draw the (1¢). Write how many there are in all.

1. How much do you spend for both?

____5____ ¢

2. How much do you pay altogether?

_____ ¢

▶ Quiz
···

Use to show each price.

Draw the (1¢). Write how many there are in all.

3. How much do you spend for both?

 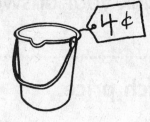

_____ ¢

4. How much do you pay in all?

_____ ¢

IPS2 Intervention • Problem Solving

Name _____   Use with Lesson 2.9.

Problem Solving Strategy 2:
Make a Picture

4 bees are on a flower.
1 flies away.
How many are left?

bee

UNDERSTAND

- Underline the question.
- Circle what you know.

PLAN

Draw ● to show the bees.

SOLVE

Draw 4 ●.   Cross out 1.

So, there are __3__ bees left.

CHECK

- How does your picture help you solve this problem?

▶ **Try These** ....

Use ● to subtract. Draw the ●.
Write the difference.

1. Lee sees 5 ants. 2 crawl away. How many ants are left?

ant   ____ ants

Intervention • Problem Solving **IPS3**

PRACTICE ON YOUR OWN

Use ● to subtract. Draw the ●.
Write the difference.

1. Pam finds 3 bugs.
   2 fly away.
   How many bugs are left?

    bug
    __1__ bug

2. Bill sees 5 birds in a tree.
   3 fly away.
   How many birds are there now?

    bird
   ___ birds

3. A cat had 6 kittens.
   2 kittens were given away. How many kittens are left?

    kitten
   ___ kittens

▶ Quiz

Use ● to subtract. Draw the ●.
Write the difference.

4. Jon has 4 worms.
   1 wiggles away.
   How many worms does Jon have now?

    worm
   ___ worms

5. Rae sees 4 moths.
   2 fly away.
   How many moths are left?

    moth
   ___ moths

Name _____   Use with Lesson 3.4.

Problem Solving Strategy 3:
Make a Picture

There are 2 plates.
There are 3 buns on each plate.
How many buns are there?

UNDERSTAND

- Draw a line under the question.
- Circle what you know.

PLAN

Make a picture. Write an addition sentence. Solve.

SOLVE

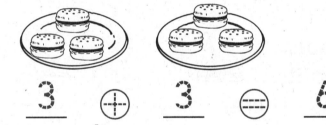

3 ⊕ 3 ⊜ 6

So, there are 6 buns.

CHECK

- How do you know that your answer makes sense?

▶ **Try These**

Make a picture to solve the problem.
Write an addition sentence to check.

1. There are 4 gray fish.
   There are 3 white fish.
   How many fish are there?

    fish

Intervention • Problem Solving IPS5

PRACTICE ON YOUR OWN

**Make a picture to solve the problem.
Write an addition sentence to check.**

1. There are 6 big sharks.
   There are 2 small sharks.
   How many sharks are there?

   __6__ ⊕ __2__ ⊖ __8__ sharks

2. There are 2 pails.
   There are 2 shells in each pail.
   How many shells are there?
   ___ ◯ ___ ◯ ___ shells

3. There are 3 red boats.
   There are 5 blue boats.
   How many boats are there?
   ___ ◯ ___ ◯ ___ boats

▶ **Quiz**

**Make a picture to solve the problem.
Write an addition sentence to check.**

4. There are 4 fast turtles.
   There are 5 slow turtles.
   How many turtles are there?
   ___ ◯ ___ ◯ ___ turtles

5. There are 2 large crabs.
   There are 4 baby crabs.
   How many crabs are there?
   ___ ◯ ___ ◯ ___ crabs

Problem Solving Strategy 4: Use Objects

5 children play.
Then 2 more come.
How many children play in all?

UNDERSTAND

- What do you need to find out?
- Circle the question.

PLAN

You can use objects to solve the problem.

Use and to show the groups of children.

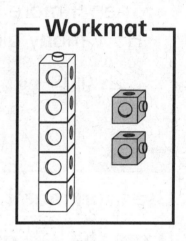

SOLVE

There are 5 white cubes and 2 gray cubes. I count 7 cubes in all.

So, there are __7__ children in all.

CHECK

- Does your answer make sense? Explain.

▶ Try These

Use Workmat 1, , and to solve.

Draw the and you use.

1. There are 2 boys playing tag.
   Then 3 girls join them.
   How many children play tag in all?

   ____ children

Name _____

PRACTICE ON YOUR OWN

Use Workmat 1, ▢ and ▣ to solve.
Draw the ▢ and ▣ you use.

1. There are 4 children running.
   There are 6 children swinging?
   How many children are there in all?
   __10__ children

2. There are 5 girls playing jacks.
   Then 4 more girls join them.
   How many girls are playing jacks in all?
   ____ girls

▶ **Quiz**

Use Workmat 1, ▢ and ▣ to solve.
Draw the ▢ and ▣ you use.

3. Amy sees 7 cats.
   Then she sees 3 more.
   How many cats does she see in all?
   ____ cats

4. There are 7 children outside.
   Then 2 more children join them.
   How many children are there now?
   ____ children

IPS8 Intervention • Problem Solving

Name _____   Use with Lesson 5.4.

Problem Solving Strategy 5: Make a Picture

There were 5 kites in the air.
Some kites landed.
2 kites are still in the air.
How many kites landed?

UNDERSTAND

- Circle the question.
- Tell what you know.

__5__ kites were in the air   __2__ kites left in the air

Think: Look at the picture. It shows how many kites are left in the air.

PLAN

You can draw a picture to solve the problem.

SOLVE

What number should I add to 2 to get 5?

$5 - \blacksquare = 2$

$2 + \underline{3} = 5$

So, __3__ kites landed.

CHECK

- How do you know that your answer is correct?

▶ **Try These**

Make a picture to solve the problem.

What number do I add to 1 to get 3?

1. At first, 3 bees were flying.
   Some landed on a flower.
   Only 1 bee is left in the air.
   How many bees landed?

   $3 - \blacksquare = 1$

   $1 + \underline{2} = 3$

   __2__ bees

Name _____

PRACTICE ON YOUR OWN

Make a picture to solve the problem.

1. There were 4 moths in the air.
   Some landed on a light.
   2 moths were left in the air.
   How many moths landed?
   __2__ moths

2. There were 6 jets flying.
   Some jets landed on a field.
   There are 4 jets still in the air.
   How many jets landed?
   _____ jets

▶ Quiz ··

Make a picture to solve the problem.

3. There were 5 birds flying.
   Some landed on a tree.
   There were 3 birds left in the air.
   How many birds landed?
   _____ birds

4. There are 10 butterflies flying.
   Some land on a fence.
   7 butterflies are left in the air.
   How many butterflies land?
   _____ butterflies

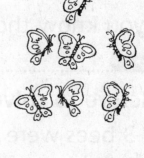

IPS10 Intervention • Problem Solving

Name _____   Use with Lesson 6.5.

Problem Solving Skill 6: Choose the Operation

Jen has 8 apples.
She gives 2 away.
How many apples
does she have left?

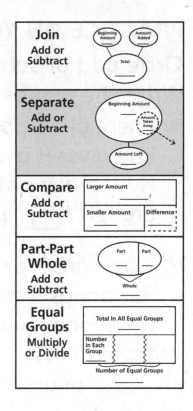

UNDERSTAND

- Underline the question.
- Circle what you know.

PLAN

Use the separate Math Map to find the amount left. Write a number sentence to solve.

SOLVE

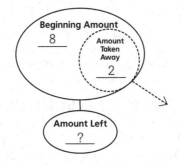

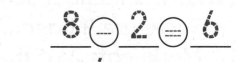

So, there are __6__ apples left.

CHECK

- How do you know you chose the correct Math Map?

▶ **Try These**

Circle add or subtract. Write the number sentence.

1. Tom has 6 carrots.
   He eats 3 of them.
   How many are left?

   __3__ carrots

        add   subtract

Intervention • Problem Solving **IPS11**

Name _____

PRACTICE ON YOUR OWN

Circle add or subtract.
Write the number sentence.

1. There are 8 potatoes.
   Rae gives 4 away.
   How many potatoes
   are left? __4__ potatoes

   Think: Do I add or subtract?

   add   subtract
   ___ ○ ___ ○ ___

2. There are 2 pears.
   Pat brings 4 more.
   How many pears are
   there now?
   _____ pears

   Think: Do I add or subtract?

   add   subtract
   ___ ○ ___ ○ ___

▶ Quiz ....

Circle add or subtract.
Write the number sentence.

3. There are 9 bananas.
   Maria eats 2 of them.
   How many bananas
   are left? _____ bananas

   Think: Do I add or subtract?

   add   subtract
   ___ ○ ___ ○ ___

4. There are 4 eggs.
   A hen lays 5 more.
   How many eggs are
   there now?
   _____ eggs

   Think: Do I add or subtract?

   add   subtract
   ___ ○ ___ ○ ___

IPS12 Intervention • Problem Solving

Name _____   Use with Lesson 7.7.

Problem Solving Strategy 7: Make a Graph

Which ice cream is the favorite of the most children?

Favorite Ice Cream	
Chocolate	||||	
Vanilla	|||| |||
Strawberry	||||

UNDERSTAND
- Draw a line under the question.
- Circle what you know.

PLAN
You can make a bar graph.

SOLVE
Count the number of votes for each flavor. Then draw a bar on the graph to represent that number.

Favorite Ice Cream

(Bar graph: Chocolate = 6, Vanilla = 8, Strawberry = 4; x-axis: Number of Votes 0–8; y-axis: Flavor)

The longest bar is for vanilla, so ___vanilla___ is the favorite ice cream.

CHECK
- How can you check that your answer is correct?

▶ **Try These**

Use your graph to solve.

1. Which flavor was the least favorite?

_____

Intervention • Problem Solving IPS13

Name _____

PRACTICE ON YOUR OWN
Make a bar graph from the data in the tally chart.

Favorite Pet	
Hamster	||||
Cat	||||
Dog	|||| ||||

Favorite Pet

Pet: Hamster / Cat / Dog

Number of Votes: 0 1 2 3 4 5 6 7 8 9 10

Use your graph to solve.

1. Circle the pet that was chosen the most.

    hamster   cat   dog

▶ **Quiz**

Use your graph to solve.

2. Circle the pet that was chosen the least.

    hamster   cat   dog

3. How many children like dogs or cats?

    ____ children

IPS14 Intervention • Problem Solving

Name _____   Use with Lesson 8.6.

Problem Solving Strategy 8: Use Objects

Joe put pieces of candy in 4 groups of ten. He has 2 pieces left. How many pieces of candy does Joe have?

UNDERSTAND

- Underline the question.
- Circle what you know.

PLAN

How will you solve this problem?
Use objects to show tens and ones.

SOLVE

Use ▭▭▭▭▭▭▭▭▭▭ ▭▭ .
Write how many tens and ones.

__4__ tens __2__ ones = __42__

So, Joe has __42__ pieces of candy.

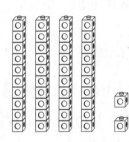

CHECK

- How do you know that your answer is correct?

▶ **Try These**

Use Workmat 3 and ▭▭▭▭▭▭▭▭▭▭ ▭▭ .
Write how many tens and ones.

1. Emma puts grapes in 2 groups of ten. She has 6 left. How many grapes does Emma have?

___ tens ___ ones = ___   Emma has ___ grapes.

Intervention • Problem Solving **IPS15**

PRACTICE ON YOUR OWN

Use Workmat 3 and ▦ ▫.
Write how many tens and ones.

1. Kevin puts dominoes in 1 group of ten. He has 5 left. How many dominoes does Kevin have?

    ___ ten ___ ones = ___

    Kevin has ___ dominoes.

2. Isabella puts barretts in 3 groups of ten. She has 3 left. How many barretts does Isabella have?

    ___ tens ___ ones = ___

    Isabella has ___ barretts.

▶ **Quiz**

Use Workmat 3 and ▦ ▫.
Write how many tens and ones.

1. Karen puts acorns in 7 groups of ten. She has 5 left. How many acorns does Karen have?

    ___ tens ___ ones = ___

    Karen has ___ acorns.

IPS16 Intervention • Problem Solving

Name _____   Use with Lesson 9.9.

Problem Solving Skill 9: Use a Model

Dan hit 25 balls. Chris hit 10 more than Dan. How many balls did Chris hit?

UNDERSTAND

Underline the question.
Circle what you know.

PLAN

Use 🧊🧊 to show the balls.

SOLVE

To find 10 more, add 1 ten.
Draw the ten.
How many cubes in all? __35__
So, Chris hit __35__ balls.

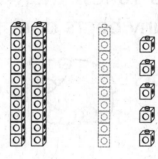

CHECK

- How do you know that your answer makes sense?

▶ Try These

Use the model. 🧊🧊
Find 10 more or 10 less.

Think: To find 10 less cross out 1 ten.

1. Phil has 31 boats. Fred has 10 less. How many boats does Fred have?

   ___ boats

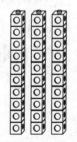

Intervention • Problem Solving IPS17

Name _____

PRACTICE ON YOUR OWN

Use the model.
Find 10 more than or 10 less than.

1. Ann has 24 stickers.
   Pete has 10 more than Ann.
   How many stickers does Pete have?

   Think: To find 10 more add 1 ten.

   __34__ stickers

2. Jeff has 32 bears.
   Nina has 10 less than Jeff.
   How many bears does Nina have?

   ____ bears

   Think: To find 10 less cross out 1 ten.

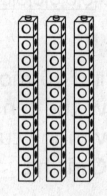

▶ Quiz ....

Use the model.
Find 10 more than or 10 less than.

3. Lynn has 50 marbles.
   Rico has 10 more than Lynn. How many marbles does Rico have?

   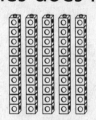

   ____ marbles

4. Juan has 13 planes.
   Scott has 10 less than Juan. How many planes does Scott have?

   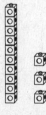

   ____ planes

IPS18 Intervention • Problem Solving

Name _____   Use with Lesson 10.7.

Problem Solving Strategy 10: Use Objects

Patrick has 9 apples. He has 3 horses. He wants to give each horse the same number of apples. How many apples will each horse get?

UNDERSTAND
- Underline the question.
- Circle what you know.

PLAN

Use ◯. Put the ◯ into equal groups.

SOLVE

Draw 3 groups for the 3 horses.
Put one ◯ into each group until all the ◯ are used up.
So, each horse gets __3__ apples.

CHECK
- Does your answer make sense? Explain.

▶ **Try These**

Use ◯ to solve. Draw the ◯.

1. The farm has 8 cows and 2 barns. The same number of cows sleep in each barn. How many cows sleep in each? _____ cows

Intervention • Problem Solving **IPS19**

Name _____

PRACTICE ON YOUR OWN

Use ○ to solve. Draw the ○.

1. Chris picked 10 ears of corn. He put them in 2 baskets. Each basket has the same amount. How many ears of corn are in each basket? _____ ears

2. Yvette has 6 pigs. She puts them in 3 pens. She puts the same number in each pen. How many pigs are in each pen? _____ pigs

▶ Quiz

Use ○ to solve. Draw the ○.

3. Dominic has 8 tomatoes. 2 tomatoes grew on each tomato plant. How many tomato plants did Dominic grow? _____ plants

4. Liza has 10 eggs. She has 5 chickens. Each chicken laid the same number of eggs. How many eggs did each chicken lay? _____ eggs

IPS20 Intervention • Problem Solving

Name _____ Use with Lesson 11.4.

Problem Solving Strategy 11: Write a Number Sentence

4 children ride bikes. 3 more come.
How many children ride bikes now?

UNDERSTAND

- Underline the question.
- Circle what you know.

PLAN

Write a number sentence to solve.

SOLVE

__4__ ⊕ __3__ ⊖ __7__ children

So, __7__ children ride bikes now.

CHECK

- How could a picture help you tell if your answer make sense?

▶ **Try These**

Write a number sentence to solve.

1. Mike has 3 cats. Then he gets 2 more. How many cats does he have now?

   ___ cats

Intervention • Problem Solving **IPS21**

Name _____

PRACTICE ON YOUR OWN

Write a number sentence to solve.

1. Liz has 8 rockets.
   Bill gives her 4 more.
   How many rockets
   does Liz have now?
   _____ rockets

   Think: Where do I put the numbers in my sentence?

   __8__ ⊕ __4__ ⊜ __12__

2. Nate counts 10 stars.
   Sam counts 3 stars. How
   many more stars does
   Nate count than Sam?
   _____ stars

   ____ ◯ ____ ◯ ____

▶ **Quiz** ....

Write a number sentence to solve.

3. Sandy paints 9 suns.
   Then she paints 2 more.
   How many suns does
   she paint in all?
   _____ suns

   ____ ◯ ____ ◯ ____

4. The children saw 12 clouds.
   Then 3 blew away. How
   many clouds do they
   see now?
   _____ clouds

   ____ ◯ ____ ◯ ____

IPS22 Intervention • Problem Solving

Name _____   Use with Lesson 12.5.

Problem Solving Skill 12: Choose the Operation

Joy has 5 dolls. Carolyn has 4 dolls. How many dolls do they have in all?

UNDERSTAND
- Underline the question.
- Circle what you know.

PLAN

You want to find how many dolls **in all**. So you need to **add**.

SOLVE

Circle add or subtract.
Write the number sentence.

(add)   subtract

$5 \oplus 4 \ominus 9$

So, they have __9__ dolls in all.

CHECK
- How do you know your answer is correct?

▶ **Try These**

Circle add or subtract.
Write the number sentence.

1. Ken has 7 airplanes. He gives 3 to Bill. How many does Bill have now?

    __4__ airplanes

add   (subtract)

$7 \ominus 3 \ominus 4$

Intervention • Problem Solving   IPS23

Name _____

PRACTICE ON YOUR OWN

Circle add or subtract.
Write the number sentence.

1 Jason has 14 cupcakes. His family eats 9. How many cupcakes are left?
   _5_ cupcakes

Think: Do I add or subtract?

add   (subtract)

_14_ ⊖ _9_ ⊜ _5_

2 Marta sets out 6 juice boxes. Carl sets out 3 more. How many juice boxes are there in all?
   ____ juice boxes

Think: Do I add or subtract?

add   subtract

____ ◯ ____ ◯ ____

▶ **Quiz**

Circle add or subtract.
Write the number sentence.

3 There are 5 spoons. Michael brings 4 more. How many spoons are there now?
   ____ spoons

Think: Do I add or subtract?

add   subtract

____ ◯ ____ ◯ ____

4 Amanda eats 12 raisins. Melissa eats 9 raisins. How many more raisins did Amanda eat?
   ____ raisins

Think: Do I add or subtract?

add   subtract

____ ◯ ____ ◯ ____

IPS24 Intervention • Problem Solving

Name _____ Use with Lesson 13.7.

Problem Solving Strategy 13: Use Objects

How many △ make a ◇ ?

UNDERSTAND

- Underline the question.
- Circle what you know.

PLAN

Use pattern blocks to make a model.

Use △ to cover the ◇.

SOLVE

Draw the shapes to show what you made.

Count the △ you see.

Write how many △ you drew.

So, it took __2__ △ to make a ◇.

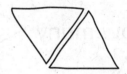

CHECK

- How do you know that your answer is correct?

▶ **Try These**

Use pattern blocks to solve.
Draw to show what you make.
Write how many pattern blocks you use.

Think: Cover the larger shape with the smaller ones.

1. How many △ make a ▱ ?

___ △

Intervention • Problem Solving **IPS25**

Name _____

PRACTICE ON YOUR OWN

Use pattern blocks to solve.
Draw to show what you make.
Write how many pattern blocks you use.

Think: Cover the larger shape with the smaller ones.

1. How many △ make a ⬡ ?

   _2_ △

2. How many ◇ make a ⬡ ?

   ___ ◇

▶ Quiz

3. How many ◇ and △ make a ▱ ?

   ___ ◇ and ___ △

   OR

   ___ ◇ and ___ △

4. How many ▱ and △ make a ⬡ ?

   ___ ▱ and ___ △

   OR

   ___ ▱ and ___ △

   OR

   ___ ▱ and ___ △

IPS26 Intervention • Problem Solving

Name _____

Use with Lesson 14.6.

Problem Solving Skill 14: Use Data from a Table

This table tells how many animals the children saw in camp.

How many raccoons and deer did the children see?

Animals Seen in Camp	
raccoons	2
ducks	5
deer	3

UNDERSTAND
- Underline the question.
- Circle what you know.

PLAN
Add the number of raccoons and deer in the table together.

SOLVE

__2__ raccoons __3__ deer   __2__ ⊕ __3__ ⊜ __5__

So, the children saw __5__ raccoons and deer in all.

CHECK
- How do you know that your answer makes sense?

▶ Try These

Use the table to answer the question.

Think: Find the numbers and subtract.

1. How many more ducks than raccoons did they see?

    __3__ more ducks   __5__ ⊖ __2__ ⊜ __3__

Intervention • Problem Solving IPS27

Name _____

PRACTICE ON YOUR OWN

This table tells how many insects the children found.

Use the table to answer the questions.

Insects Found		
crickets		2
ants		5
beetles		3
butterflies		6

1. How many beetles and ants did they find?
   __3__ beetles and ants

   Think: Find the numbers and add.

   _3_ ⊕ _5_ ⊜ _8_

2. How many more butterflies than crickets did they find?
   _____ more butterflies

   Think: Add the numbers and subtract.

   ____ ○ ____ ○ ____

3. How many insects did they find in all that were not butterflies?
   _____ insects

   Think: Add the numbers for insects that are not butterflies.

   ____ ○ ____ ○ ____

▶ Quiz ....

Use the table to answer the questions.

4. How many more ants than beetles did they find?
   _____ more ants

   Think: Find the numbers and subtract.

   ____ ○ ____ ○ ____

5. How many insects did they find in all that were not ants?
   _____ insects

   Think: Find the numbers for insects that are not ants and add.

   ____ ○ ____ ○ ____

Use with Lesson 15.5.

Problem Solving Skill 15: Choose a Method

Scott ate 12 grapes. Tal ate 4. How many did they eat in all?

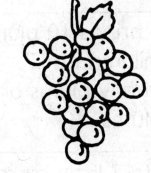

UNDERSTAND
- Underline the question.
- Circle what you know.

PLAN
You can use mental math, paper and pencil, or ▭▭▭ to solve the problem.

SOLVE

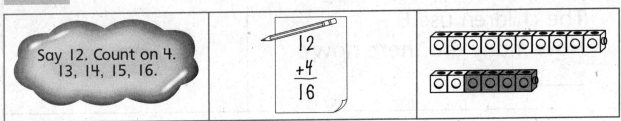

| Say 12. Count on 4. 13, 14, 15, 16. | 12 +4 ---- 16 | (cubes) |

So, the boys ate __16__ grapes in all.

CHECK
- Which way would you use to solve this problem? Why?

▶ **Try These**

Choose the best way to solve the problem.

1. There are 14 bees on the basket. Then 4 fly away. How many bees are left?

   __10__ bees

Intervention • Problem Solving IPS29

Name _____

PRACTICE ON YOUR OWN

Choose the best way to solve the problem.

1. Maria brings 10 plums. The children eat 7. How many plums are left? __3__ plums	
2. Rob ate 7 little carrots. Jean also ate 7. How many did they eat in all? _____ carrots	
3. There are 12 napkins. The children use 8. How many are there now? _____ napkins	

▶ **Quiz** ....

Choose the best way to solve the problem.

4. There are 9 hot dogs. Dad cooks 5 more. How many are there now? _____ hot dogs	
5. Mom baked 15 cookies. Grandma baked 9. How many more did Mom bake? _____ cookies	

IPS30 Intervention • Problem Solving

Name _____ Use with Lesson 16.6.

Problem Solving Strategy 16: Use Objects

10 girls are sledding. 5 more come. How many girls are sledding?

UNDERSTAND

- Underline the question.
- Circle what you know.

PLAN

Use ☐ and ■ to show the girls.

SOLVE

Begin with 10 ☐.

Next, count 5 more ■.
Then, count all the tiles.

So, 15 girls are sledding.

CHECK

Does your answer make sense? Explain.

▶ **Try These**

Use Workmat 1, ☐, and ■.

Draw the ☐ and the ■ you use.

1. The class made 12 snowmen. 9 are still here. How many snowmen melted?
_____ snowmen

Intervention • Problem Solving IPS31

Name _____

PRACTICE ON YOUR OWN

Use Workmat 1, ☐, and ■.

Draw the ☐ and the ■ you use.

1. 6 boys are ice skating. There are 13 boys in all. How many boys are not ice skating?
   __7__ boys

   Think: I will sh
6 tiles. How m
more tiles do I r
to get to 13

2. There are 8 children in the fort. 5 more are behind trees. How many children are there in all?
   _____ children

▶ Quiz

Use Workmat 1, ☐, and ■.

Draw the ☐ and the ■ you use.

3. Lou sees 11 children sledding. Then 4 more come. How many children are sledding?
   _____ children

4. 12 mittens are missing. The children find 7. How many mittens are still missing?
   _____ mittens

Name _____ Use with Lesson 17.5.

Problem Solving Strategy 17: Act It Out

Sam cuts a pie into 4 equal parts. What does Sam's pie look like?

UNDERSTAND

- Underline the question.
- Circle what you know.

PLAN

Act it out using fraction circles.

SOLVE

Find fraction circle pieces that you can use to make a complete circle with 4 equal parts. Draw the pie.

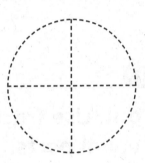

CHECK

- Does your answer make sense? Explain.

▶ **Try These**

Draw equal parts. Act it out. Use fraction circles or bars.

1. Jack cuts a pie into 3 equal parts. What does his pie look like?

2. Sarah cut a granola bar in half. What does her granola bar look like?

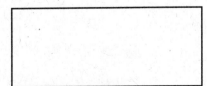

Intervention • Problem Solving IPS33

PRACTICE ON YOUR OWN

Act it out. Use fraction circles or bars. Draw equal parts.

1. Kim cuts a cake into 4 parts. The parts are equal. What does her cake look like?

2. Jim and 3 other boys share a brownie. The parts are equal. What does their brownie look like?

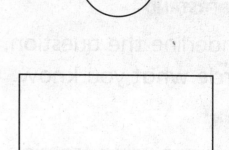

Quiz

Act it out. Use fraction circles or bars. Draw equal parts.

3. Rita and Lee share a pie. They will each eat one half. What does their pie look like?

4. Jane, Fay, and Nate share a pie. They each get an equal part. What does their pie look like?

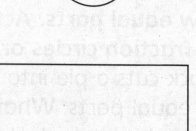

IPS34 Intervention • Problem Solving

Name _____   Use with Lesson 18.7.

Problem Solving Strategy 18: Act It Out

Ron wants to buy a top for 6¢. What ways can he use 🪙 and 🪙 to make 6¢?

UNDERSTAND

- Underline the question.
- Circle what you know.

PLAN

Act it out. Use coins.

SOLVE

Find two ways.

Nickels	Pennies	Total Value
5	1	6¢
0	6	6¢

CHECK

- How can you check to see if your answer is correct?

▶ **Try These**

Nan wants to buy a pen for 5¢. What ways can she use and to make 5¢?
List how many coins you choose to total 5¢.

Ways to Make 5¢	
Nickels	Pennies
1. Think: Try 0	5
2.	

Intervention • Problem Solving IPS35

PRACTICE ON YOUR OWN

Mia wants to buy a doll for 10¢. What ways can she use 🪙, 🪙, and 🪙 to make 10¢?

Show 3 ways to make 10¢.
Use 🪙, 🪙, and 🪙.

	Ways to Make 10¢		
	Dimes	Nickels	Pennies
1	1	0	0
2			
3			

▶ **Quiz**

Nick wants to buy a book for 20¢. What ways can he use , , and 🪙 to make 20¢?

Show 3 ways to make 20¢.
Use 🪙, 🪙, and 🪙.

	Ways to Make 20¢		
	Dimes	Nickels	Pennies
4			
5			
6			

Name _____   Use with Lesson 19.5.

Problem Solving Strategy 19: Act It Out

What coins could you use to buy these two things?

UNDERSTAND

- Underline the question.
- Circle what you know.

PLAN

Use coins to act this problem out.

SOLVE

So, you can use 3 _____ to show the total amount.

CHECK

- How do you know the coins you used are correct?

▶ **Try These** ....

Draw the coins you would use to buy the items.

1.

Intervention • Problem Solving IPS37

Name _____

PRACTICE ON YOUR OWN

Draw the coins you would use to buy the items.

Think: Use a quarter, a nickel, and pennies.

▶ **Quiz** ....

Draw the coins you would use to buy the items.

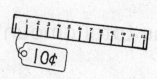 

IPS38 Intervention • Problem Solving

Name _____

Use with Lesson 20.8.

Problem Solving Skill 20:
Use Data From a Graph

Which mealtime is the favorite of the most classmates?

UNDERSTAND

- Underline the question.

PLAN

Use the graph to answer the questions.

SOLVE

<u>3</u> classmates chose breakfast.

<u>4</u> classmates chose lunch. <u>2</u> classmates chose dinner.

So, <u>lunch</u> is the favorite meal.

CHECK

- How did you know lunch was the favorite meal?

▶ **Try These**

Use the graph to answer the questions.

1. Which day did more classmates choose? Circle it.
   (Saturday)   Sunday

2. How many more children chose Saturday than Sunday?
   _____ children

Intervention • Problem Solving **IPS39**

Name _____

PRACTICE ON YOUR OWN
Use the graph to answer the questions.

1. Which color was chosen by the fewest children? Circle it.

    red
    yellow
    green
    blue

2. Which color was chosen the most? Circle it.

    red
    yellow
    green
    blue

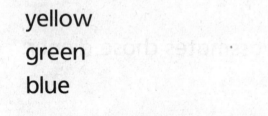

3. How many children like red or blue?

    _____ children

▶ Quiz

Use the picture graph to answer these questions.

4. How many more children chose blue than green?

    _____ children

5. How many like green or yellow?

    _____ children

Use with Lesson 21.5.

Problem Solving Skill 21: Make Reasonable Estimates

Lynn makes necklaces. About how many beads long is the string?

about 1   about 3   about 5

UNDERSTAND

- Underline the question.
- Circle what you know.

PLAN

See which estimate makes sense.

SOLVE

Picture the beads on the string as you study the estimates.

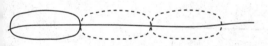

Think: About 1 bead is too few. About 5 beads are too many.

So, about __3__ beads long makes sense.

CHECK

- Does your answer make sense? Explain.

▶ Try These

1. About how many beads long is the string? Circle the answer that makes sense.

   Think: Draw the beads on the string to help you estimate.

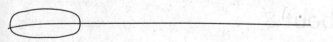

   about 4   about 6   about 8

Intervention • Problem Solving IPS41

Name _____

PRACTICE ON YOUR OWN

About how many beads long is the string?
Circle the answer that makes sense.

1.

about 2   about 5   (about 8)

Think: In your mind, try to see the beads on the string.

2.

about 1   about 3   about 6

▶ Quiz

About how many beads long is the string?
Circle the answer that makes sense.

3.

about 2   about 4   about 7

4.

about 3   about 5   about 7

5.

about 3   about 6   about 8

Name _____   Use with Lesson 22.4.

Problem Solving Strategy 22: Use Objects

Which object is heavier?
Which object is lighter?

UNDERSTAND

- Underline the questions.
- Circle what you know.

PLAN

Use a and ⬚ to weigh each object.

SOLVE

The marker weighs __7__ ⬚.

The tape dispenser weighs __25__ ⬚.

The __tape dispenser__ is heavier,

and the __marker__ is lighter.

CHECK

- How could you check that your answer is correct?

▶ **Try These**

Use a and ⬚ to weigh each object.
Circle the heavier object.

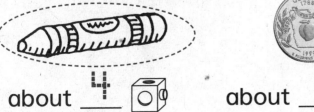

about __4__ ⬚   about __1__ ⬚

Intervention • Problem Solving IPS43

Name _____

PRACTICE ON YOUR OWN

Use a ⚖ and 🧊 to weigh each object. Circle the heaviest object. Mark an X on the lightest object.

1

about ____ 🧊   about ____ 🧊   about ____ 🧊

▶ Quiz

Use a ⚖ and 🧊 to weigh each object. Circle the heaviest object. Mark an X on the lightest object.

2

   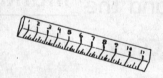

about ____ 🧊   about ____   about ____ 🧊

Name _____   Use with Lesson 23.4.

Problem Solving Skill 23: Choose the Measuring Tool

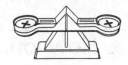

Mia wants to find out how much her juice box holds. Which tool should she use?

UNDERSTAND

- Underline the question.
- Circle what you know.

PLAN

Study the tools. Think about what they measure.

SOLVE

 A measures how long or wide.

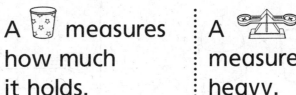

A measures how much it holds.

A measures how heavy.

So, Mia will use a _____.

CHECK

- Mia would use a _____ to see if it is heavier than her apple.

▶ **Try These**

Circle the correct tool to measure.

1. How wide is it?

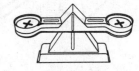

PRACTICE ON YOUR OWN

Circle the correct tool to measure.

1. How much will it hold?

 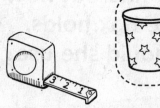  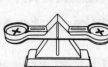

2. How long is the ribbon?

  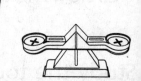

3. Which is heavier?

  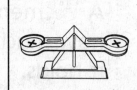 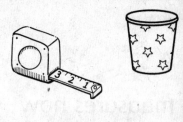

▶ **Quiz**

Circle the correct tool to measure.

4. How tall is the desk?

 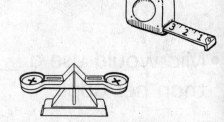

5. Which box is heavier?

  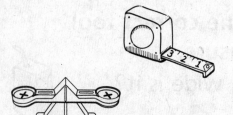

Name _____   Use with Lesson 24.7.

Problem Solving Skill 24: Choose a Method

There are 50 ants on a cracker.
10 crawl away.
How many ants are left?

UNDERSTAND

- Underline the question.
- Circle what you know.

PLAN

You can use mental math, paper and pencil, or a calculator to solve the problem.

SOLVE

You can use mental math.	You can use paper and pencil.	You can use a calculator.
5 tens minus 1 ten equals 4 tens. The answer is 40. _____ ants	50 -10 —— 40	40

CHECK

- Which other way could you use to solve this problem?

▶ **Try These**

Choose a way to solve the problem.

1. There are 27 ladybugs on the flowers. 20 more join them. How many ladybugs are there in all?

   47 ladybugs

Intervention • Problem Solving IPS47

Name _____

PRACTICE ON YOUR OWN

Choose a way to solve the problem.

1 Carson saw 20 green caterpillars and 30 brown caterpillars. How many did he see in all? **50** caterpillars	
2 There are 60 bananas on the tree. 13 fall off. How many bananas are left? _____ bananas	
3 33 orange butterflies fly into the garden. 12 yellow butterflies join them. How many are there in all? _____ butterflies	

▶ **Quiz** ....

4 There are 26 birds singing in a tree. 13 fly away. How many are left? _____ birds	
5 There are 54 snakes in the zoo. 22 more join them. How many snakes are there? _____ snakes	

IPS48 Intervention • Problem Solving

Name _____

Problem Bank

1. Bill has 2 pens. Then he gets 3 more. How many pens does he have in all?

   ____ pens

2.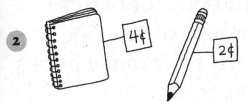

   How much do they cost in all?

   ____ ¢

3. Jan sees 4 dogs. Then 2 run away. How many dogs are left?

   ____ dogs

4. Li feeds 3 cats. Then 4 more come. How many cats are there now?

   ____ cats

5. There are 9 logs. Ron uses some to make a fire. There are 4 left. How many logs did Ron use?

   ____ logs

6. Ms. Lee has 6 balls. She lends some to children. There are 3 left. How many balls does Ms. Lee lend?

   ____ balls

7. There are 4 bins. There are 2 books in each bin. How many books are there in all?

   ____ books

8. There are 7 bees on a bush. Then 2 more land there. How many bees are there in all?

   ____ bees

9. There are 4 men in a van. Then 3 more get in. How many men are in the van now?

   ____ men

Intervention • Problem Solving IPS49

Name _____

10. Ted saw 8 crows in the air. Some landed on a fence. There are 3 crows still in the air. How many crows landed?

____ crows

11. There are 7 buns. Pete eats 3 of them. How many buns are left?

____ buns

For 12–13, use the graph.

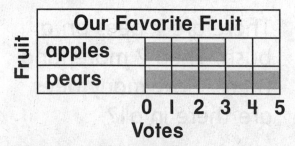

12. Did more children choose pears or apples?

_____

13. How many more children chose pears than apples?

____ more

14. Kira picks 4 apples. Then she picks 5 more. How many apples does she pick in all?

____ apples

15. Sarah puts books in 2 groups of ten. She has 3 left. How many books does Sarah have?

____ books

16. Mark puts stamps in 4 groups of ten. He has 9 left. How many stamps does Mark have?

____ stamps

17. Roy has 33 marbles. José has 10 marbles less than Roy. How many marbles does José have?

____ marbles

18. Maria has 48 rocks. Paula has 10 more than Maria. How many rocks does Paula have?

____ rocks

IPS50 Intervention • Problem Solving

Name _____

19. There are 5 cookies on each plate. How many cookies are there on 2 plates?

____ cookies

20. There are 10 apples in each bag. How many apples are in 3 bags?

____ apples

21. There are 11 paper cups. Mom uses 5. How many cups are left?

____ cups

22. A cook bakes 8 cakes. Then he bakes 4 more. How many cakes does he bake in all?

____ cakes

23. Fran moves 3 chairs. Mike also moves 3 chairs. How many chairs do they move in all?

____ chairs

24. Use ▱ and △ to create a picture

25. Draw a boy **to the left of** a tree.

26. Draw a fence **near** a house.

Intervention • Problem Solving IPS51

Name _____

For 27–28, use the table.

Birds at the Feeder	
Robins	4
Jays	2
Wrens	5

27 How many robins and wrens were at the feeder?

_____ robins and wrens

28 How many more robins than jays were at the feeder?

_____ more robins

29 Liz has 18 books. She lends 4. How many books does Liz have left?

_____ books

30 Dan eats 10 grapes. Then he eats 8 more. How many grapes does he eat in all?

_____ grapes

31 Margo peels 5 apples. She needs to peel 12 apples in all. How many more apples does she need to peel?

_____ apples

32 Lena picks 8 carrots. Dirk picks 8 more carrots. How many carrots do they pick in all?

_____ carrots

IPS52 Intervention • Problem Solving

Name _____

33 There are 7 ducks swimming. There are 15 ducks in all. How many ducks are not swimming?

____ ducks

34 Hank cuts a cookie. It has 4 parts. The parts are not equal. Which cookie is his?

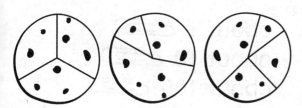

35 Rose and Phil share a cookie. They will each eat one half. Which is their cookie?

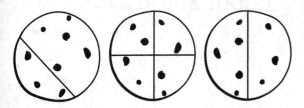

36 Eric wants to buy a card for 12¢. What ways can he use , , and 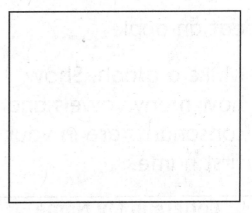 to make 12¢? Draw the coins.

37 Draw the coins for each amount.

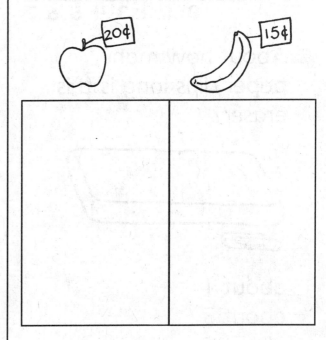

Name _____

38 Which takes a shorter amount of time?

sweep a room

open a door

39 Which takes a longer amount of time?

watch a movie

eat an apple

40 Make a graph. Show how many vowels and consonants are in your first name.

Letters in My Name							
Vowels							
Consonants							

0 1 2 3 4 5 6 7

41 About how many paper clips long is this eraser?

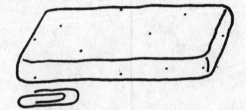

about 1
about 3
about 5

42 About how many paper clips long is this pencil?

about 2
about 4
about 10

43 About how many 🖇 does a pen weigh?

about 1 🖇
about 10 🖇
about 100 🖇

44 About how many 🖇 does a penny weigh?

about 3 🖇
about 30 🖇
about 300 🖇

IPS54 Intervention • Problem Solving

Name _____

45 Eve buys a new book. She wants to find out how heavy it is. What measuring tool can she use to find this out?

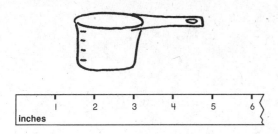

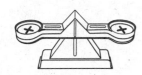

46 Rob finds his baby shoe. He wants to find out how long it is. What measuring tool can he use to find this out?

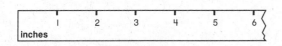

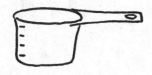

47 Ellen has 60 stickers. She gives 20 away. How many stickers are left?

_____ stickers

48 Ian sees 40 ants. Max sees 10 more. How many ants do they see in all?

_____ ants

49 Beth buys juice for 15¢. She buys a bagel for 22¢. How much money does she spend in all?

_____ ¢

50 Brandon has 75¢. He buys a toy sailboat for 32¢. How much money does he have left?

_____ ¢